SIMPLE MACHINES ALL AROUND US

LEVERS

AT WORK

BY MAGGIE S. WESLEY

Enslow PUBLISHING

DISCOVER!

Please visit our website, www.enslow.com. For a free color catalog of all our high-quality books, call toll free 1-800-398-2504 or fax 1-877-980-4454.

Library of Congress Cataloging-in-Publication Data

Names: Wesley, Maggie S., author.
Title: Levers at work / Maggie S. Wesley.
Description: New York : Enslow Publishing, [2022] | Series: Simple machines all around us | Includes index.
Identifiers: LCCN 2020013397 | ISBN 9781978520844 (library binding) | ISBN 9781978520820 (paperback) | ISBN 9781978520837 (set) | ISBN 9781978520851 (ebook)
Subjects: LCSH: Levers–Juvenile literature.
Classification: LCC TJ207 .W47 2022 | DDC 621.8–dc23
LC record available at https://lccn.loc.gov/2020013397

Published in 2022 by
Enslow Publishing
101 West 23rd Street, Suite #240
New York, NY 10011

Designer: Katelyn E. Reynolds
Editor: Kristen Nelson

Photo credits: Cover, pp. 1 ttsz/ iStock / Getty Images Plus; cover, pp. 1–24 (background and typeface) Reytr/Shutterstock.com; cover, pp. 1–24 (paper) jannoon028/Shutterstock.com; cover, pp. 1–24 (note sticker) Kindlena/Shutterstock.com; p. 5 aldomurillo/ E+/Getty Images; pp. 7, 10 VectorMine/ iStock / Getty Images Plus; p. 9 Nasky/Shutterstock.com; p. 11 Thanit Weerawan/ Moment/Getty Images; p. 13 stocknroll/E+/Getty Images; p. 15 aywan88/ iStock / Getty Images Plus; p. 16 Universal History Archive/Getty Images; p. 17 AlinaMD/ iStock / Getty Images Plus; p. 19 Solidago/E+/Getty Images; p. 21 (salad tongs) TanawatPontchour/ iStock / Getty Images Plus; p. 21 (nut cracker) hudiemm/E+/Getty Images; p. 21 (tweezers) AntonioGuillem/ iStock / Getty Images Plus; p. 21 (hammer) IJdema/ iStock / Getty Images Plus; p. 21 (pliers) Srinrat Wuttichaikitcharoen / EyeEm/ Getty Images; p. 21 (broom) 4x6/ iStock / Getty Images Plus.

Portions of this work were originally authored by Gillian Houghton Gosman and published as *Levers in Action*. All new material this edition authored by Maggie S. Wesley.

Printed in the United States of America

Some of the images in this book illustrate individuals who are models. The depictions do not imply actual situations or events.

CPSIA compliance information: Batch #CSENS22: For further information contact Enslow Publishing, New York, New York, at 1-800-398-2504.

Contents

Boldface words appear in Words to Know.

LEVERS IN LIFE

Seesaws aren't just fun to play on. They're also important machines for doing work. A seesaw is a simple machine called a lever. Levers are boards or beams that lay on **fulcrums**. Simple machines are tools that make work easier by changing certain forces.

LEVERS, LIKE THE OTHER FIVE KINDS OF SIMPLE MACHINES, DON'T HAVE MANY PARTS.

How Levers Work

Levers are useful simple machines because they help people lift heavy loads! Here's how: When a force, or effort, is applied to one end of the board, the fulcrum allows the other end to overcome **resistance** and lift a weight.

THE AMOUNT OF WEIGHT A LEVER CAN LIFT IS PART OF ITS **MECHANICAL ADVANTAGE.**

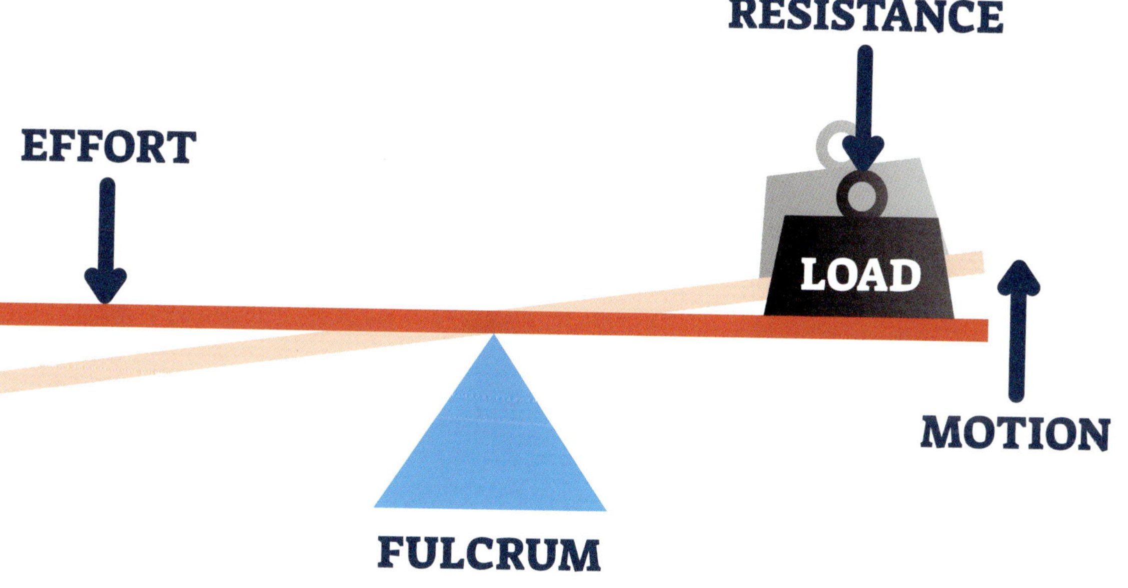

Three Cool Classes

There are three classes, or kinds, of levers. What makes them different is where the fulcrum is placed. The closer the fulcrum is to the end of the lever holding up the load, the easier the load is to lift.

THE LONGER A LEVER'S HANDLE, THE MORE **LEVERAGE** IT HAS.

FIRST CLASS LEVER

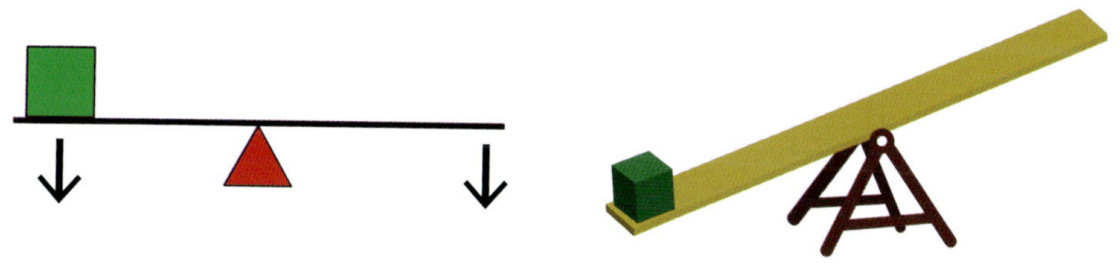

SECOND CLASS LEVER

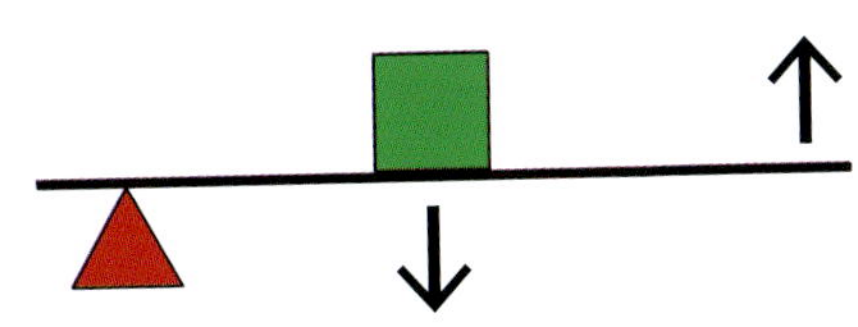

THIRD CLASS LEVER

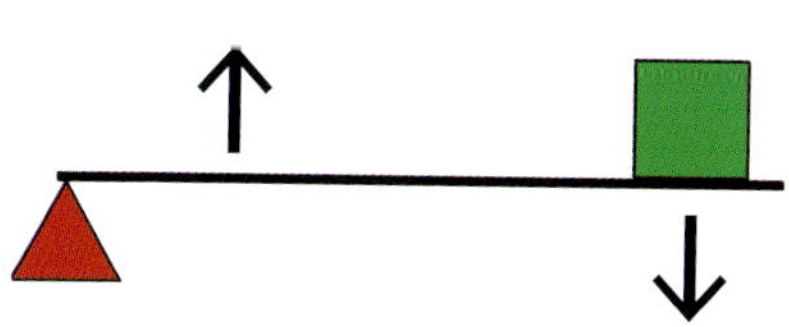

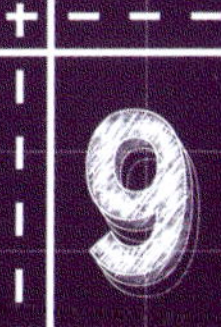

Fantastic First Class Levers

The fulcrum sits about directly between the two ends of a first class lever. **Scissors** are two first class levers put together. For both scissor levers, effort is applied to one end, where you hold on. A mid-board fulcrum allows the other end to overcome resistance.

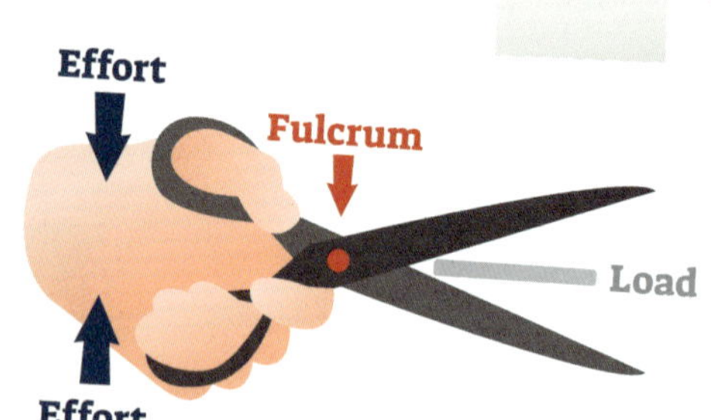

THE FORCE YOU APPLY AND THE FORCE YOU MAKE ARE AT OPPOSITE ENDS OF A FIRST CLASS LEVER.

Sensational Second Class Levers

A second class lever has a fulcrum at one end of the board and the resistance that's overcome is in the middle. A wheelbarrow is an example. The wheel at one end is the fulcrum and the load is in the center.

THE FORCE YOU CREATE WITH A WHEELBARROW'S HANDLES OVERCOMES A LOAD'S RESISTANCE.

Terrific Third Class Levers

A third class lever is like a second class lever because the fulcrum is at one end, but the force you apply and the force you make are **switched**. Your arm is a third class lever when you lift something.

YOUR ARM **MUSCLE** IS THE FORCE YOU APPLY WHEN YOUR ARM ACTS AS A THIRD CLASS LEVER.

The History of Levers

Around 1500 BCE, people started using a lever called a shaduf (or shadoof) to water fields. A shaduf has a weight at the end of the board that is near the fulcrum. The weight helps raise water collected in a pail bound to the board's longer end.

FARMERS STILL USE SHADUFS IN INDIA AND EGYPT TO WATER THEIR FIELDS.

Modern Levers

Greek inventor Archimedes discovered a lever's mechanical advantage around the third century BCE. Today, levers are used in many **compound machines**. Oil pump jacks are one example. They pull up oil from the ground using a long beam that rests on a tall fulcrum.

AN **ENGINE** MAKES THE FORCE AN OIL JACK PUMP'S LEVER USES.

Find Levers Around You

Now that you know what levers are, let's see how many you can spot! You know scissors use two levers, but did you know **tweezers** and nail clippers are also levers? Levers, like other simple machines, will continue to be important in our everyday lives.

HOW MANY OF THE LEVERS PICTURED HERE CAN YOU FIND AROUND YOUR HOME?

THIRD CLASS LEVER

SALAD TONGS

NUT CRACKER

SECOND CLASS LEVER

THIRD CLASS LEVER

TWEEZERS

FIRST CLASS LEVER

HAMMER PULLING OUT A NAIL

PLIERS

FIRST CLASS LEVER

THIRD CLASS LEVER

SWEEPING WITH A BROOM

Words to Know

compound machine Two or more simple machines that work together to make a task or job easier.

engine A machine that makes power.

fulcrum The point on which a lever turns around.

leverage The added help of using a machine to do work.

mechanical advantage The advantage, or benefit, gained from using a machine or tool to change a natural force.

muscle One of the parts of the body that allow movement.

resistance A force that goes against the direction of a moving object, slowing it down.

scissors A tool made from two blades joined in the middle, with the sharp sides facing each other, that's used for cutting paper and other thin materials.

switch To replace or change something with something else.

tweezers A small tool made from two thin pieces of metal joined at one end that's used to pull, move, and hold small objects.

For More Information

Books

Crane, Cody. *Simple Machines*. North Mankato, MN: Children's Press, 2019.

Rivera, Andrea. *Levers*. Minneapolis, MN: Abdo Zoom, 2017.

Spilsbury, Louise. *Levers*. New York, NY: PowerKids Press, 2019.

Websites

Dirtmeister's Science Reporters: Simple Machines, Lever
teacher.scholastic.com/dirtrep/simple/lever.htm
Learn more about the forces that make levers work!

Lever
www.khanacademy.org/science/physics/discoveries/simple-machines-explorations/a/lever
Instructional videos on this site can help you understand why levers make work easier.

Simple Machines | Science Trek
ny.pbslearningmedia.org/resource/idptv11.sci.phys.maf.d4ksim/simple-machines/
Check out a cool video and learn more about the simple machines in your daily life!

Index